Bayron Ruiz

FORESTRY HARVESTING AND MANAGEMENT ON 100 ha OF TROPICAL MOIST FOREST

Bayron Ruiz

FORESTRY HARVESTING AND MANAGEMENT ON 100 ha OF TROPICAL MOIST FOREST

Forest harvesting and management in the tropics

ScienciaScripts

Imprint
Any brand names and product names mentioned in this book are subject to trademark, brand or patent protection and are trademarks or registered trademarks of their respective holders. The use of brand names, product names, common names, trade names, product descriptions etc. even without a particular marking in this work is in no way to be construed to mean that such names may be regarded as unrestricted in respect of trademark and brand protection legislation and could thus be used by anyone.

Cover image: www.ingimage.com

This book is a translation from the original published under ISBN 978-620-0-00933-3.

Publisher:
Sciencia Scripts
is a trademark of
Dodo Books Indian Ocean Ltd. and OmniScriptum S.R.L publishing group

120 High Road, East Finchley, London, N2 9ED, United Kingdom
Str. Armeneasca 28/1, office 1, Chisinau MD-2012, Republic of Moldova, Europe
Managing Directors: Ieva Konstantinova, Victoria Ursu
info@omniscriptum.com

Printed at: see last page
ISBN: 978-620-8-50350-5

FORESTRY HARVESTING AND MANAGEMENT ON 100 ha OF TROPICAL MOIST FOREST

DR. BAYRON ALEXANDER RUIZ BLANDON

TABLE OF CONTENTS

1. EXECUTIVE OVERVIEW

Since ancient times, man has continuously used wood as a material for the construction of tools and structures, and for the production of by-products that have been and are important for the development of society (latex, pulp mills, medicine).

The intensive use of forest resources has had and continues to have detrimental effects on the environment; there is a drastic reduction of green areas worldwide due to the exploitation of forests and jungles without applying reforestation programmes; there are also areas that are impossible to recover, and there are species of fauna and flora that are on the verge of extinction and are already extinct.

Most forest harvesting is for industrial purposes, some for ornamental purposes, although these are becoming increasingly important, and very little is based on the maintenance of the natural environment.

By elaborating a Systematic Random Inventory in the municipality of Medio Baudó, in the village of Pie de Pepé (Quebradas Pepé Claro and Pepé Sucio), the aim is to programme a Forest Management Plan that will reduce the negative impacts generated by indiscriminate logging and the irrational use of both renewable and non-renewable natural resources. The natural regeneration species will also be quantified and qualified in order to know their abundance, diversity and the type of vegetation cover existing in this forest for its future management, conservation or exploitation. Sustainable forest management seeks to guarantee the permanence of forested areas in terms of their extension, composition and characteristics, allowing forest management and harvesting to be carried out without significantly reducing the economic possibility of permanent production of goods and services, and to conserve the stability of the natural ecosystem, biodiversity and forest heritage.

The Forest Harvesting Plan is the description of the systems, methods and equipment to be used in the harvesting of the forest and extraction of the products, presented by the interested party in carrying out single forest harvests. The Forest Management Plan is the formulation and description of the silvicultural systems and work to be applied in the forests subject to harvesting, in order to ensure their sustainability.

Forest Use and Management Plans allow us to know the structure of the vegetation, the richness of the species and the state of the forest, that is to say, to obtain qualitative and quantitative floristic information on the forest, which facilitates making decisions on the use or sustainable management of natural tropical forests, in order to use renewable and non-renewable natural resources efficiently and rationally.

The sustainable use and management of forests implies that all the products available within the ecosystem are well known, in order to better plan their use, to know their economic potentialities in a comprehensive way and to have several economic use alternatives available, in order to make the use that is most advisable based on the biological and silvicultural particularities of the forest.

The aim is to carry out a Forest Harvesting and Management Plan in a forest area of 100 hectares in the department of Chocó, municipality of Medio Baudó, in the village of Pie de Pepé (Quebradas Pepe Claro and Pepe Sucio). In addition, to carry out a statistical inventory in a forest area of 100 ha, with a sampling intensity of 2%, i.e. 2 ha; to obtain information on the species that make up the forests; to calculate and analyse the conventional indices for the species found in the forest inventory, in the sampling area; to determine the total existing volume of standing timber and harvestable timber per unit area in the sampling area; describe the possible ecological and social impacts that may be caused or produced by forest harvesting in the study area; and carry out descriptive studies of the biophysical and socio-economic conditions of the community of Pie de Pepé (Quebrada Pepe Claro and Pepe Sucio), municipality of Medio Baudó.

2.RATIONALE AND STRATEGIES FOR THE SUSTAINABLE USE, MANAGEMENT AND HARVESTING OF NATURAL FORESTS IN THE BAUDO ENVIRONMENT

2.1SOCIAL BASE.

The municipality of Medio Baudó, exactly the village of Pie de Pepe, is very welcoming, due to the quality of its inhabitants, their kindness and cordiality. Public services such as energy, aqueduct, health and education are not efficient and have low coverage rates, it has a health centre and the level of education is basic primary and secondary education. There is also a police station and churches. The population of the village of Pie de Pepé is the most organised. The community manifests that groups and associations have been created in different ways around different topics of interest. Perhaps one of the most important indicators of their organisational capacity is the creation of the community radio station, through which important bonds of solidarity have been established between the community of Pie de Pepé and the neighbouring settlements. Another of the current associations in Pie de Pepé is the ASPROCAPEB Association, as well as other associations of artisans and folkloric groups.

2.2ECONOMIC BASIS

The economy of the village of Pie de Pepe is based on agriculture, artisanal mining and logging; most of the income of the families in this village comes from these activities. practices. In addition, hunting of animals in the forest is carried out, mainly for subsistence purposes.

2.3 ENVIRONMENTAL BASE

The environmental part offers the possibility of protecting the natural resources, the environment and biodiversity, which form the basis for a reasonable use and preservation of the biotic and abiotic potential of the area, generating a better way of life for the inhabitants of this community.

3.GENERAL INFORMATION

3.1Time of use.

3.2GEOGRAPHICAL AND ADMINISTRATIVE LOCATION.

The municipality of Medio Baudó is located in the central part of the department of Chocó. Its municipal seat is Puerto Meluck, a town located on the left bank of the Baudó River at 05°11'66.5" north latitude and 76°57'28.7" west latitude of the Greenwich meridian, at a distance of approximately 95 km from Quibdó.

The municipality has an area of 1,390.6 km^2 and is bordered to the north by the municipality of Alto Baudó, to the south by the municipality of Bajo Baudó, to the east by the municipality of Istmina and the municipality of Cantón de San Pablo, and to the west by the municipality of Alto Baudó. The Pie de Pepé district has an area of 8,374.74 ha, which is equivalent to 6.01% of the total area of the municipality of Medio Baudó. The Pie de Pepe district is bordered to the north by the municipality of Istmina, to the south by the district of Beriguadó, to the west by the district of Corundó and to the east by the municipality of Istmina. Pie de Pepé is located at 100m above sea level; its rural villages are Pie de Pepé, Boca de Berrecuí and Aguacatico.

3.3IDENTIFICATION OF THE APPLICANT AND TECHNICAL MANAGER

3.3.1Applicant. Luís Ernesto Mosquera "Presidente Consejo Comunitario del corregimiento de Pie de Pepé".

3.3.2Technical Manager. Engineer Ascanio Arriaga Arango.

3.4DESCRIPTION OF THE FOREST MANAGEMENT UNIT.

3.4.1 Area and Boundaries. The village of Pie de Pepé is part of the Baudó river basin, which is part of the tropical rainforest macro-ecosystem of the Pacific. Most of its territory is flat and jungle-like, with an average altitude of 13 metres above sea level.

The Baudó River runs from south to north through a narrow valley that is gradually widening. This is the natural and transport axis of the district and the municipality through which the populations are integrated and communicated.

3.4.2 Ownership and Acquired Rights. Afro-Colombian communities own land by traditional family inheritance. Each village owns a territory as a community and within this, each family has its own land for agricultural and livestock work. The use of forest areas is done individually or collectively with the permission of the community. The Afro-Colombian communities, as ethnic groups with a cultural tradition, are covered by Law 70/93, for which the state is obliged to title the lands they have traditionally occupied collectively. The ACABA community council, based on this law, has presented a request to INCORA for collective titling of the territory corresponding to the middle and upper Baudó basin and part of Pie de Pepe. The requested territory corresponds to uncultivated land located within the area delimited as a forest reserve by Law 2 of 1959.

Within the area covered by the application there are some lands already titled by INCODER, which means that they would be excluded from the collective titling. The application covers the entire territory of the middle Baudó and part of the lower and upper Baudó.

During 1999, a new community council was formed that includes part of the communities of Pie de Pepé, from Angostura to Berrecuí. This Community Council has made its respective request for titling, separating itself from the ACABA titling. Currently, the processing of these applications is in the process of study and re-definition, as some communities do not agree and INCORA considers that a single titling of a territory that covers more than one municipality is inconvenient.

There are several indigenous communities settled in this area, of which eight (8) were incorporated in the departmental ordinance. And six (6) are legally recognised as resguardo according to INCODER's register of indigenous communities. The indigenous communities located in this area have the corresponding collective titles that accredit them as autonomous territorial entities within the municipality. These are the Resguardo del Río Torreidó - Chimaní, the Resguardo Santa Cecilia, the Resguardo Puerto Libre, the Resguardo Trapiche, the Resguardo Purricha and the Resguardo Dabeiba - Querasito.

3.4.3 History of the forest. The forest has been harvested for both commercial and domestic use by the communities, with selective extraction of species according to the tastes and preferences of the community, according to the quality and commercial value of the wood. In addition, the respective request to the competent entity is not made, nor are there any plans for use and management that guarantee the normal and sustainable yield of the forest.

3.5SOCIAL CONTEXT

The corregimiento of Pie de Pepé is made up of a system of small settlements; most of the territory is covered with forests and jungles. Pie de Pepé has semi-urban characteristics, with a population of 783 inhabitants. Until 2003, the urban centre of Boca de Pepé was the municipal capital of Medio Baudó, and after the issuance of Ordinance No. 008 of 2003, the town of Puerto Meluck was established as the main administrative centre of the municipality.

3.5.1 Socio-economic and demographic aspects. The location of the settlements on the municipal territory is mainly concentrated in six axes that correspond to the most important water elements; a central axis on the Baudó River, secondary axial axes formed by the Pepé River, the Berreberre River, the Misará and Torreidó Rivers, the Baudocito River and tertiary axes formed by tributary streams where some populations are settled. The Puerto Meluck - Pie de Pepé road will constitute a new axis which will gradually gain momentum as mobility conditions improve with the construction of the bank and the increase in the frequency of transport. The village of Pie de Pepé has 783 inhabitants, with a population density of 27.6 inhabitants/ha. There are 425 concentrated dwellings and 25 dispersed dwellings, for a total of 450 dwellings in this territory.

3.5.2 Public services. The low coverage and poor quality of public services were identified by the community as the two most important issues. factors that have the greatest impact on the low quality of life offered by the corregimiento.

3.5.2.1 Telecommunications. In this village the communication service is deficient, with a telephone line that functions as a receiver (COMPARTEL). This situation is due to the non-existence of a contract between Telecom and the community, as well as some money that the community owes to this entity, after a bad administration of the service, which is used by the whole community, it does not totally supply the needs of the population. The television and radio services have low coverage and are regular. It currently operates in a community hut.

3.5.2.2 Education. The educational facilities in the corregimiento are limited to schools; the only school that functions does not have its own facilities. In relation to the coverage and quality indicators, the following was determined:

The level of coverage of pre-school, primary and secondary education varies, with secondary being the lowest at only 6%, followed by pre-school at 45%; primary figures show 100% coverage. The service does not cover the total demand of the community (see fig. 1). Buildings are of fair quality, lack sanitation facilities and adequate sports facilities. The schools are poorly and insufficiently equipped.

3.5.2.3 Health. The village has an infrastructure of a health post administered with resources from the mayor's office. The health service has very low coverage in terms of medical personnel, as there is only one professional who attends, and there is no support network of nurses or health promoters who attend the health post.
The health post has an acceptable physical infrastructure, but it is in need of repair, as the buildings are in poor condition. The quality of the equipment is also deficient, as there is at least one stretcher.

3.5.2.3 Water supply. The coverage of the service is ample, since most of the population has access to it on a daily basis. The quality of the water supply is not good due to the lack of treatment of the water supply and the absence of storage tanks.

3.6 LOCAL ECONOMY.

For the analysis of the local economy in the corregimiento, the economic activities were divided into primary, secondary and tertiary sectors, without losing the interrelation that exists between them as productive chains.

3.6.1 Primary Sector. The primary sector is the economic production that depends on the immediate exploitation of natural resources. This category of production includes agriculture, livestock and fish farming, mining and timber extraction.

3.6.1.1 Agriculture. Although the agricultural sector is the basis of the municipality's economy, its productivity and commercialisation is incipient, with the production of pan coger predominating throughout the territory. A few surpluses are marketed in Istmina or Buenaventura. According to the survey to producers, it was established that out of the 19 respondents, only 20% of the products are marketed.

According to the ACABA document, the form of agricultural exploitation is totally artisanal and traditional, basically using "the socola, tumba y pudre technique, which has its own particularities for each product, but in general consists of opening up a plot of forest that is primary or that has been resting for three to five years; To do this, first the bushes, palms and small trees are cut with a rula to proceed to water the seed on them, and then the larger trees are felled with an axe so that their decomposition can be used as fertiliser'. The extension that each producer cultivates varies between three and five hectares; it is estimated that each family needs at least six hectares to survive, two of which are dedicated to bananas. In order to facilitate rotation and according to the availability of labour, they usually have several plots in different stretches of the river and of different qualities. in order to have banana crops and at least one annual harvest of maize and rice. Each plot is used for a maximum of two harvests and then left to rest.

a) Location

Agricultural production takes place throughout the territory, because it is the economic base of all its inhabitants. Pie de Pepé is one of the municipalities of Medio Baudó with the greatest extension of cultivated land.
The main crops grown in this area are: bananas, plantains, maize, rice, sugar cane, chontaduro and yucca. (See fig. 4). Plantain and maize are the products that occupy most of the territory. It is important to highlight that given the form of exploitation, technically classified as agroforestry and artisanal transitory crops, they do not affect the forest ecosystem to a great extent.

3.6.1.2 Livestock Production. Livestock production in the corregimiento is incipient and mostly for self-consumption.

The breeding of the pigs is done without any technique, because once the sow brings out the young, the offspring are raised next to the sow without any care. The owner is only concerned with feeding the female; once the offspring are able to defend themselves, they are taken to the maize fields to grow between harvests. Other animal species raised are chickens and ducks, but in small quantities and only for family consumption. At Pie de Pepé, experimental breeding of some species of fish is underway.

3.6.1.3 Mining. It is not a significant activity in this community because, although there is knowledge of the existence of precious metals in the Pepé River, no production figures are known. This is done in an artisanal manner using the mazamorreo technique.

3.6.1.4 Fishing. It is not an economically profitable activity for the population, because the few species that can be found there are on the verge of extinction. The little fishing activity is done for family consumption. Most of the fish consumed is from the sea and comes from the coasts of Pizarro.

3.6.15 Forest exploitation. One of the most important economic activities in Pie de Pepé is the indiscriminate exploitation of timber, therefore, it is difficult to quantify, but it is important to highlight that it is the most traded product with Buenaventura and the department of Risaralda.
The most felled or harvested species are Cedar, Oak, Ceiba, Abarco, Peine mono, Sande, Chibuga, Guino and Pantano.

3.6.2 Secondary Sector. The secondary sector is understood as all economic activities that involve the processing of raw materials for the generation of industrial inputs or final products for consumption. In this sector of the economy, in Pie de Pepé, the following are found manufacturing industries and food processing for consumption.

3.6.2.1 Small Industry (Manufacturing). There is a small-scale manufacturing industry in this town. At the mouth of Pepé, there is a small mattress industry that is taking its first steps without sufficient marketing and sales volumes. There is also a bakery that has been in sufficient demand that it is expected to grow.

There is also a sawmill that produces a good amount of wood per year when weather conditions permit. This is located in the ravine, Boca de Pepé.

3.6.3 Tertiary Sector.

3.6.3.1 Trade. Bearing in mind that this is a particularly agricultural village and that its products, as mentioned above, are mostly for internal consumption and the surplus production is sold to intermediaries who sell in the municipalities of Istmina and Bajo Baudó, it can be foreseen that there is also a low level of income from this concept. The little trade that does exist is made up of small food and grocery shops.

Consumption of manufactured goods and processed food is purchased in Buenaventura, Istmina and Quibdó, according to the survey of producers.

3.8 HOUSEHOLD.

In most of the forests of this community, wildlife extraction or hunting and community fishing is practised, generally comprising hilly and mountainous areas where a great variety of wildlife persists, which serves as the basis for the population's diet.

3.9 AGRICULTURE. Traditional agriculture is practised in this area, represented in agroforestry and miscellaneous systems of crops and transitional crops, combined with the extractive use of medicinal plants and the use of natural vegetation. The predominant crops are bananas, maize, rice, cassava, barajó and fruit trees, among others. The use of agriculture has different modalities in the corregimiento:

3.9.1 Agroforestry Systems. When simultaneously on a plot or parcel of land a crop qualified as agricultural is planted, whether temporary, semi-permanent or permanent, and is interspersed with trees for the production of wood, firewood, fruit, resins or secondary products in long periods of production. It can also be combined with pasture for livestock (silvopastoral) or a combination of all three (agroforestry). In this area this system is used with permanent or semi-permanent crops with six-monthly or annual cycles such as bananas, Borojó and Chontaduro. In addition to forest species such as cedar.

3.9.2 Miscellaneous agriculture. This type of activity is mainly carried out with permanent and/or transitory intermingled crops of different species, generally of bread crops. The size of the plots ranges from small to medium (0.5 to 20 hectares). Especially banana, cassava and fruit trees.

3.9.3 Transitional Agriculture. This activity consists of the opening or felling of areas of forest in which a wide variety of temporary crops are sown in small areas of fallow relict forest, paddocks in association with small plots of maize and rice crops.

3.9.4 Livestock. Cattle ranching is a use that is just being introduced in this corregimiento, and it is done on a small scale. It is important to take into account that, although this use is not yet significant, it tends to grow and in some cases it is being carried out in unsuitable areas.

3.9.5 Mining. This is not a very common activity in the municipality. There are restricted areas in the Pepé river basin, where artisanal mining with low impact on the water sources and the associated hydrophane is carried out, although the most frequent practice is the mazamorreo.

3.10 FOREST EXTRACTION.

Forestry activity is currently not very widespread, but it represents an important use within the zone, as a large part of the total area is dedicated to this activity. This extraction consists of selective thinning of commercially valuable species, which results in a balance of species diversity and density of the forest harvested; this practice is carried out in the upper zones of the rivers, with emphasis on the Pepé river basin.

3.11 NON-TIMBER PRODUCTS.

The inhabitants of this community do not attach much importance to non-timber forest products, i.e. all products other than timber that are extracted from the forest, due to a lack of knowledge about their use, but in some cases medicinal and aromatic plants are extracted for commercial purposes.

3.12 ENVIRONMENTAL SERVICES.

These services are essential for the good development of the community, such as renewable natural resources like fauna, flora, microbiological resources, soil and non-renewable resources that are of vital importance for the subsistence of this community.

3.13 CULTURAL SERVICES.

These are important for the community because livelihoods, customs and beliefs are inherited from generation to generation, preserving traditional and ancestral traits. The festivities of the Virgen del Carmen are held on 16 July.

3.14 GEOMORPHOLOGY, TOPOGRAPHY AND SOILS.

Forest soils under tropical rainforest (bh-T) are characterised by the fact that they have developed under the influence of a forest canopy, with a marked effect of a shallow root system, the association of organisms, and the presence of a forest canopy. specific vegetation and soil and litter layer, or litter together with nutrient washing. There is a close interrelation of chemical, physical and biological properties in association with climatic conditions of high temperature and high rainfall in the soil formation factors. According to the Use Capability Classification System, the soils of Pie de Pepé are classified into eight and designated by

Roman numerals which represent groups of soils that have the same relative degree of risks or limitations in their use, which become progressively larger from class I to VIII.
Soils of the first four classes are capable of producing crops under appropriate management conditions. Soils of classes V, VI and VII are suitable for the use of adaptable native plants; classes V and VI can produce specialised crops and ornamentals. Class VIII soils are not suitable for agricultural activity.
Subclasses are groups of capacity units within classes, which have the same degree of dominant constraints for agricultural use. The following are recognised within the subclass level
Four constraints: erosion (e); moisture, drainage and flooding (h); root zone constraints (s); and climatic constraints (c).
The soils of the corregimiento of Pie de Pepé have potential uses that include various types of agriculture, forestry and conservation of fauna and flora, such as land suitable for agriculture, land suitable for permanent improved crops, land suitable for forestry and land suitable for wildlife.

3.15 HYDROGRAPHY AND HYDROLOGY

The high precipitation determines a dense drainage pattern, consisting of numerous streams (rivers and streams) and reservoirs (swamps). The hydrographic cycles of the watercourses are regulated by precipitation. The various watercourses are of great importance in the area, as there are no other means of communication. The water system of the municipality consists of a system of rivers and streams that are tributaries of the Baudó River. The Baudó basin is the third most important water system in Chocó, after the Atrato and the San Juan, and is comparable to the latter in terms of flow velocity, but of lesser importance in terms of geographical location.

The Pepé River is the most important tributary of the Baudó River, navigable for a large part of its course, especially during the winter season. It rises to the northwest of Bocas de Pepé and has an approximate length of 24km. The most important tributaries are: Quebrada la brea, Beriguadó and Sandó.

3.16 CLIMATE

The district has the following Climatic Unit:

Warm Humid and Perhumid (Cp), covering 100 % of the territory.

The climate of the area is determined by:

- Offshore winds that circulate from the ocean to the continent.
- Orographic conformation of the area: The western mountain range and its foothills impede the passage of northern winds, thus contributing to the high precipitation recorded in this area; in addition, its location in the intertropical zone of the equatorial calms, with low atmospheric pressure, high cloudiness and constant temperature, allow the formation of different microclimates.
- It is important to highlight the influence that the Humboldt Current has on the climate of the region by modifying the temperature of the south-easterly trade winds as they pass through the current.

With the help of the information obtained from the IDEAM, it will allow a more concrete analysis of the climatic behaviour of the corregimiento, through data processed with precise statistical methods, which will be related to the cartographic and mapped information of the different parameters to be analysed.

On the other hand, representative stations for the municipal area were evaluated, from which six stations were selected. The stations provided rainfall records, while the synoptic station recorded information on temperature, relative humidity, sunshine and cloudiness.

3.16.1 Precipitation. Pie de Pepé is subject to a bimodal rainfall regime, product, and permanent precipitation values.

high due to the regional confluence of humid air masses from the indicated origins. With annual rainfall between 4,000 and 10,000 mm, heavy rainfall occurs almost daily for most of the year. The climatic conditions of the area are high temperatures, humid air, and abundant rainfall, 7,500 mm on average.

This climate generates unhealthy conditions due to the humidity and muddy conditions, which creates breeding grounds for mosquitoes. It has dry periods during the months of December, January, February and March; winter in April, May, June, September to November; in July and August there is a transition period.

3.16.2 Temperature. Air temperature is a variable of minor seasonal variation, the changes from month to month in the same place are fundamentally due to differential states of atmospheric cloudiness and therefore to variations in the incidental radiation at the terrestrial surface, temperatures oscillate on average between 26 and 28°C. (See fig.6).

3.16.3 Relative Humidity. The average relative humidity of the stations with this information is 90% on average, both in the rainy and dry periods. However, it is important to note that towards the west it is quite high, with values above 90% (see fig. 7). In general, values are higher during the dry period, with some increases also occurring during the wet period.

3.16.4 Sunshine. The number of hours of sunshine in the area is largely influenced by precipitation in the different months of the year. In the station with heliographic recording the dry period shows the highest insolation, while the wet period records the lowest values. Sunshine values range between 65 and 110 hours per month, with November being the lowest and February the highest.

3.16.5 Wind speed. Wind speed is relatively low with a daily average of 1.0 to 1.5 metres per second (-m/s-), with a bimodal distribution with the Intertropical Confluence Zone; relative maxima occur in the months of April, May and November. The diurnal variation of wind speed coincides, in general terms, with what normally occurs in the tropical region. Wind speeds are highest in the midday hours, intermediate in the early evening and lowest in the early morning.

3.16.6 Evaporation. Evaporation is defined as the loss of water from a soil completely covered by a low green crop, through soil evaporation and plant transpiration without water limitation. The analysis of evapotranspiration synthesises the climate, as it integrates several atmospheric elements and serves as a basis for applied research.

such as water requirements for irrigation (water balances and calculation of indices), which serve to establish comparisons and concrete classifications of a climate (Holdridge, 1978).
Monthly evaporation in the Pie de Pepe district shows the following behaviour: In general, values do not vary much during the year. However, in all representative stations, the highest values are obtained between the months of March and June with records that oscillate between 125 and 140 millimetres per month. From July onwards, the values decrease very little in relation to those of the first period, with records ranging between 120 and 135 millimetres.

3.16.7 Water Balance. Based on precipitation and evaporation, the water consumption of crops can be estimated in the so-called Water Balance and the water availability of a particular area or site can be determined. In addition, periods of water deficiencies and excesses can be established over the course of the year. The water balances were calculated on the basis of monthly average and monthly precipitation data. From this comparison, rainfall excesses or deficits are defined, which are related to the useful soil reserve. For the analysis of the water balance, it is convenient that the first month of the whole series is at the end of the dry season (summer), when soil water reserves are known to be depleted. When the amount of water provided by precipitation exceeds the storage capacity of the soil, excesses are generated. Excesses of water levels are decreasing, with ranges between 300 and 900 millimetres, but are still high. Deficits occur when precipitation fails to restore the useful amount of water, in this case theoretically 100 mm.

4.FOREST MANAGEMENT UNIT PLANNING

4.1ZONING.

Based on the field work of the statistical forest inventory, it has been established that 90% of the 100 hectares of the study area are suitable for timber production, according to the needs of the population and the marketing needs of the owner of the property and the Community Council.

4.1.1Short Unit Planning

4.1.1.1 Annual Cutting Units. The 90 ha of productive forest are projected to be harvested over four years, which represents an annual cutting area of 22.5 ha during the time of implementation of the Harvesting and Management Plan. It will take into account local experience, uses, preferences and customs to initiate forest harvesting.

Ensuring that natural regeneration continues to provide a sufficient number of individuals for all species, varying the number between regrowth and regeneration. saplings, as established under current natural conditions. If all the stands found reach the stage of mature stands, taking into account their quantity, a type of management that guarantees the presence of the best and most desirable species is warranted, and if, on the other hand, given the influence of adverse factors, there is not enough regeneration to ensure the number of stands to guarantee persistent production, it will be necessary to advance enrichment programmes in the forest to ensure the number of stands to guarantee persistent production, If, on the contrary, given the influence of adverse factors, there is not sufficient regeneration to ensure the number of stands to guarantee persistent production, it will be necessary to carry out enrichment programmes in the areas with forest or to carry out reforestation in the areas where the forest has disappeared or has been impoverished due to inadequate, exaggerated and intensive use.

4.1.1.2 Extraction routes. The timber extraction routes will be aquatic and terrestrial for minor and major transport respectively.

4.1.1.3 Stockpiling site and camp. Aquatic extraction for minor transport consists of moving the logs using mules from the felling site to the banks of the river or stream; in this case the timber will be stored in storage sites. Major transport is carried out after the logs, blocks or pieces have been bundled and prepared at the storage site using trucks.

5ECOLOGICAL CHARACTERISATION

5.1FLORISTIC CHARACTERISATION

5.1.1 Forest types. According to the ecological map of Colombia, elaborated according to the ecological zoning project of the Pacific IGAC, the following vegetation cover can be distinguished in the corregimiento:

5.1.1.1 Low Altitude and Foothill Forests (Bc). Corresponds to the zonal forests, with characteristics due to the prevailing conditions; they develop in an altitudinal range from sea level to approximately 800 m above sea level and with a maximum limit of 1000 m. They are not conspicuously marked by limiting factors in their formation (flooded soils, alluvial soils). They are not conspicuously marked by limiting factors in their formation (waterlogged soils, alluvial soils).

They occupy topographical positions corresponding to colluvial-alluvial fans, hills, mountain foothills. The most representative species according to the IVI (Importance Value Index) (IGAC, 1984) are: Sande (Brosimum Utíle), Cuangare (Virola Reide), Caimito (Pouteria sp.), Nuanamo (Virola sp.), Carbonero (Hirteja racemosa), Anime (Protium sp.), Chanú (Sacoglotis procera), Guasco (Eschweilera sp.). Mora (Clarisia racemosa), Soroga (Vochysia ferruginea), Guamo or Guabo (Inga sp.), Carrá (Huberodendron patínoe), Abarco (Cariniana pyriformes), Zanca de Araña (Chrysochlamis sp.), Peine Mono (Apeaba áspera), Jigua (Ocotea sp.).

5.1.1.2 Alluvial Forests (Bb). This denomination includes a whole variety of associations whose main difference is given by the edaphic conditions that are related to the levels of flooding caused by excess runoff, which remains for periods of time ranging from hours to weeks and up to six months or almost the whole year with a sheet of water on the ground. This is how the dominance of a small number of species that are adapted to these constraints. People name them according to the species present, such as the "panganales", which correspond to what UNESCO (1973) classifies as swamp forests and the dominant species is the Raphia taedigera palm.

Some others such as Güino (Carapa guianensis), Nuanamo (Virola sp), Roble (Tabebuia rosea). The "cuangariales" classified as low altitude peaty forests (Unesco), with Cuángare (Virola sp, Otoba gracilipes), Sajo (Campnosperma panamensis), "sajales" with dominance of Sajo and Camarón (Alchornea sp). Also within this alluvial category, excellent heterogeneous forests develop under conditions of better drainage, and on terraces and fans.

5.1 FOREST INVENTORY

The statistical forest inventory carried out in the municipality of Medio Baudó, in the village of Pie de Pepé, was for the purpose of making a harvesting and management plan in accordance with the requirements of the national forestry statute and the forestry statute of CODECHOCÓ, i.e. in accordance with decree 1791 of 1996 and resolution 987 of 1998 of the corporation. There is a large area of community forest within the village that can be used for the extraction of forest products by the community's farmers. Within the area of the community forest of the PIE DE PEPE COMMUNITY COUNCIL, the evaluation or management unit was defined, consisting of a compact block of community forest of 100 ha (1,000 x 1,000m), i.e. the forest management plan was formulated within this area, to enable the community to study forest use in accordance with the regulations.

5.1.1 Characteristics of the forest sampling. In order to carry out the inventory, a 1,000 m long baseline was taken as a reference axis, drawn in an east-west direction, which divides the community forest or evaluation unit into two equal halves of 500 m x 500 m, which were called sub-block 1 (B1) and sub-block 2 (B2). -block 2 (B2). On this baseline, four inventory lines or sampling units of 500m long and 10m wide were drawn alternately in a north-south direction, clearly determining the sampling units.

Each line or strip of 500m x 10m, is divided into 10 subplots of 50m x 10m, thus constituting the registration units, which in the 4 lines give a total of 40 plots or Registration Units and within each of these registration units were measured all the trees of the different species that were found within the area with a D.A.P. greater than or equal to 10cm.

Subsequently, in order to comply with current regulations within each sampling unit, a record was made of the number of saplings (ct1), saplings (ct2), and grasses (ct3), for all species. In the 100 hectare forest area or assessment unit described in the above terms, a forest inventory was made on 2 hectares of forest corresponding to 2% of the total assessment unit, this being the sampling intensity.

5.1.2 Registration and calculation of the information. The records in the inventory were made taking the following information: number of plots, line or sampling unit; subplot number or registration unit, regional name of each one of the trees; diameter at breast height (D.A.P) of the existing trees, commercial and total height in metres of each of the trees; registration of the natural regeneration of all the species present in the sampling units taking into account the following categories (Renuevos or seedlings, when they present heights less than 30cm; Brinzal, when they present heights between 31 and 150cm; Latízales when they present heights greater than 150cm and diameters less than 9.9cm).

5.2. Calculation Method. For the analysis of the statistical information and the calculation of the commercial volumes found within the work area, the corresponding formulas were used for the elaboration of each one of them.

$AB = 0.7854 \times DAP^2$

$Vol. = AB \times hc \times Ff$

The following formulas were also used with Microsoft Excel:

Similarly, the volume table for standing trees, volume with bark of the Teresita was used.

Where:

AB	=	Basal area (m).2
DAP	=	Diameter at breast height with bark (measured at 1.30 m at ground level).
Vol.	=	Volume (m)3
he	=	Commercial height in (m)
Fait h	=	Shape factor (0,7)

❖ **Density:** The number of trees recorded per unit area or total sampling area.

$$❖\ D = \frac{\textit{Número de árboles}}{\textit{Área total de la muestra en ha}}$$

❖ **Abundance:** The number of trees per species recorded in each sampling unit. It can be absolute and relative, absolute abundance refers to the total number of individuals per species counted in the inventory, where:

❖ **Aa** = Number of individuals per species.

❖ **Relative abundance:** The percentage ratio in participation of each species to the total number of trees.

$$Ar = \frac{\textit{Número de individuos por especies x 100}}{\textit{Número de individuos en el área muestreada}}$$

❖ **Frequency:** This is the presence or absence of a species in each of the sampling units and can be absolute or relative:

◆ **Absolute frequency:** The percentage ratio ofthenumber of sampling units in which a species occurs to the total number of sampling units and its formula is as follows:

$$Fa = \frac{\textit{Número de unidades de muestreo en que ocurre una especie x 100}}{\textit{Número total de unidades de muestreo}}$$

◆ **Relative frequency**: The percentage ratio of the absolute frequency of a species to the sum total of the absolute frequencies of all species, the formula is as follows:

Absolute frequency of a species x 100

$$\text{Fr} = \frac{\textit{Frecuencia absoluta de una especie x 100}}{\textit{Suma total de frecuencia absolutas}}$$

❖ **Dominance**: This is the degree of species cover as an expression of the space occupied by the species, and can also be absolute or relative.

◆ **Absolute dominance**: Defined as the sum of the basal areas of the same species present within each sampling unit, expressed in m^2.

◆ **Relative dominance:** This is expressed as a percentage and is given by teratio between the basal area of a species and the sum total of the absolute dominances of all the species recorded in the inventory, the equation used is:

$$\text{Dr} = \frac{\textit{Área basal de cada especie x 100}}{\textit{Área basal total en el área muestreada}}$$

❖ **Importance Value Index "IVI":** This is given by the sum of the parameters expressed as a percentage of abundance, frequency and relative dominance and is used for descriptive and quantitative studies of forest type structures. **IVI = Ar% + Fr% + Dr%.**

❖ **Mixing coefficient:** This is expressed as the proportion between the number of species found by the total number of trees inventoried, the result gives a fractional that represents the average number of individuals of each species within the forest type, to calculate it the following relationship is used.

$$\text{Cm} = \frac{\textit{Número de especies}}{\textit{Número total de individuos}}$$

General **Shannon Wienner** diversity (H').

$$H' = -\sum (n_i / N) x \frac{\log_{10}\left(n_i/N\right)}{\log_{10}(2)}$$

Where:

H`: Diversity.

n_i: Number of individuals of species i. N: Total number of individuals, $\sum n_i$.

pi :

n_i Ratio of the number of individuals of species i to the number of

individuals of species $pi : \frac{n_i}{N}$ total

Log_{10}: Logarithm based on 10. Sorensen's Qualitative Coefficient (C_s).

$$C_s = \frac{2x\sum C}{\sum a + \sum b} x100$$

Where:

C_s : Sorensen's similarity coefficient.

a: Number of species in community or sample 1.

b: Number of species in the community or sample 2.

C: Number of species common or occurring in both communities.

Morisite Index (**M_i**):

$$I_m = \frac{2\sum(X_i x Y_i)}{(\lambda_1 * \lambda_2)(N_1 x N_2)}$$

$$\lambda_1 = \frac{\sum[X_i(X_i - 1)]}{N_1(N_1 - 1)}$$

$$\lambda_2 = \frac{\sum[Y_i(Y_i - 1)]}{N_2(N_2 - 1)}$$

$N_1 = \sum X_i$ $N_2 = \sum Y_i$ Where:

I_m = Morisite index.

X_i : Number of individuals of species i in community or sample 1.

Y_i . Number of individuals of species i in the community or sample 2.

N_1 : Total number of individuals of all species in community or sample 1.

N_2 : Total number of individuals of all species in the community or sample 2.

Simpson's Index (**Is**):

$$D = \frac{[n_i(n_i - 1)]}{N(N-1)}$$

$$I_s = \frac{1}{D} \qquad (0 < D < 1)$$

Where:

I_s : Simpson's Index.

D: Species diversity.

n_i : Number of individuals of species i. N: $\sum n_i$ = Total abundance of the species.

❖ **Volume**: The results of the forest inventory, carried out in the area to be managed, gave a high reliability for the volumes of the forest species of interest.

The teresita volume table was also used for the cubing of the trees.

5.3FLORISTIC RICHNESS AND DIVERSITY.

5.3.1 Species abundance. The most abundant species found in the sampling area were: the Milkvetch (Brosimum utile) with a total of 80 individuals, with a relative abundance of 6.993%, equivalent to 22% of the total number of individuals found. Palma memé (Wettinia quinaria) with 31 individuals and a relative abundance of 4.545%, representing 14% of all individuals, Hormigo (Miconia sp) with 35 trees, with a relative abundance of 3.059%, corresponding to 10% of the individuals in the sampled area, Carrá (Humberodendron patinoi) with 35 trees, a relative abundance of 3,059%, equivalent to 9% of all individuals counted, Caimito de monte and Caidita (Nectandra sp) with 31 individuals, a relative abundance of 2.710% and 1.836% respectively, equivalent to 8% each, Anime (Protium veneralense) with 30 trees, a relative abundance of 2.622%, representing 8% of all trees inventoried. Belly palm (Triartea delfoidea) with 25 individuals, a relative abundance of 2.185%, equivalent to 7% of the total number of individuals. Sangre gallina (Vismia panamensis) and Algodoncillo (Croton killipianus) with 24 individuals, a relative abundance of 2.098% for both species, equivalent to 7% of all trees inventoried.

5.3.2Frequency. The most frequently occurring species in each of the sampling units were Aceite maria (Calophyllum mariae), Algodoncillo (Croton killipianus), Anime (Protium veneralense), Aserrín (Parkia oppositifolia), Boteco (Matisia sp.), Caidita (Nectandra sp), Caimito de monte, Canelo (Licaria limbosa), Cargadero (Cymbopetalum sp), Carbonero (Licania durifolia), Carrá, (Humberodendron patinoi), Cascajero (Macrocnemum sp), Castaño (Compsoneura atopa), Cedro macho (Tapirira miriantus), Chocó, (ChoíbaDipteris panamensis), Otobo (Osteohhloem platyspermum), Palma meme (Wettinia quinaria), Palma zancona (Socrotea exorrhiza) and Palma taparo (Orbingya cuatrecasana).

5.3.3 Dominance. The most dominant species in the sampled area are the Lechero (Brosimum utile) with 8,638m^2 ; Carrá (Humberodendron patinoi) with 3,276m^2 ; Boteco (Matisia sp.) with 2,612m^2 ; Caimito de monte with 1,708m^2 ; Cedro macho (Tapirira miriantus) with 1,509m^2 ; Hormigo (Miconia sp) with 1,494m^2 ; Anime (Protium veneralense) with 1,475m^2 ; Saithe (Licania durifolia) with 1,272m^2 ; Cottonwood (Croton killipianus) with 1,228m^2 and Caidita (Nectandra sp) with 1,221m .^{2}

5.3.4 Importance Value Index. The species with the highest IVI are: Lechero (Brosimum utile) with 22.87% corresponding to 29%. Carra (Humberodendron patinoi) with 9.85% corresponding to 12%. Boteco (Matisia sp.) with 8.37% representing 10%. Palma memé (Wettinia quinaria), with 7.81%, equivalent to 9%. Caimito de monte with 6.84% or 8%. Ant (Miconia sp) with 6.83% corresponding to 8%. Anime (Protium veneralence), with 6.36% representing 7%. Cedro macho (Tapirira miriantus) with 5.54%, equivalent to 6%. Cotton grass (Algodoncillo) with 5.41%, equivalent to 6%. And Caidita with 5.14% representing 5%.

5.3.5 Characterisation of Natural Regeneration. The use, conservation and management of natural regeneration allows the sustainability of a forest, which is why it is necessary to apply silvicultural techniques and harvesting and management plans when using our forests. The best represented natural regeneration species were Cottonwood with 11 individuals, Incibe with 9 trees, Anime with 8 individuals, Palma memé and Otobo with 7 trees respectively, Lechero with 6 trees and Aceite maría with 5 trees.

5.3.6 Volume and Diametric Structure.

5.3.6.1 Volume. For diameter category **I,** 419 trees were recorded with a volume of 39,125 m^3 . In diameter category **II**, 411 individuals were found with a volume of 143,303 m^3 . In category **III**, the following were recorded 183 trees with a volume of 170,426 m^3 . For category **IV**, 82 individuals with a volume of 162,209 m^3 were recorded. In category **V**, 23 individuals were found with a volume of 83,324 m^3 . In category **VI**, 22 trees were recorded with a volume of 114,673 m^3 . For category VII, 2 trees with a volume of 12,273 m^3 were recorded, category **VIII** recorded 1 tree with a volume of 6,329 m^3 and for category **X**, 1 tree with a volume of 11,307 m^3 was recorded.

5.3.6.2 Basal area. A basal area of 5,694 m^2 was recorded for category **I;** category **II** has a basal area of 15,946 m^2 ; category **III** recorded 15,011 m^2 ; category **IV** recorded 11,237 m^2 ; category **V** recorded a basal area of 4,804 m^2 ; in category **VI** a basal area of 6,663 m^2 was recorded; for category **VII** a basal area of 0,827 m^2 was recorded; for category **VIII** a basal area of 0,503 was recorded and for category **X** a basal area of 0,785 m^2 was recorded.

5.3.6.3 Number of trees per line and area sampled. A total of 1,144 individuals were counted in the sampled area, where the most representative species was Lechero (Brosimum utile) with 80 individuals, followed by Palma memé (Gustavia sp) with 52 individuals, Carra (Humberodendro patinoi) with 35 individuals, Hormigo (Lunania paruiflora spr.et Benth) with 35 individuals, Boteco (Matisia sp) with 31 individuals, Caimito de monte (N.N) with 31 individuals, Anime (Protium nervosum cuatr)) with 30 individuals and Palma barrigona (Cyanthea spp) with 25 individuals. In terms of the number of trees per line and sampling unit, those with the highest number of individuals were: line one with 292 trees, line one with 288 trees, line three with 284 trees and line four with 200 individuals.

5.3.6.4 Species characterisation or stratification according to Ogawa. The Ogawa stratification shows a swarm of isolated points, indicating canopy gaps at intermediate levels, suggesting a number of differential strata in the forest profile.

5.3.6.5 Shannon Weiver Index (H`). This index was calculated to measure the heterogeneity or diversity in area, and for this case a value of 7.353 was obtained, indicating that there is little disturbance and high species diversity in the Pie de pepe forest.

5.3.6.6 Shannon Weiver index per family (H`). The Shannon Wiener index per family recorded a value of 0.149 indicating that the Pie de Pepe forest has little ecosystem disturbance and a high diversity of families.

5.3.6.7 Simpson's index. The sampling carried out in this research and its corresponding calculation indicate that the probability of taking two individuals at random from the same species is minimal (0.000041).

5.3.6.8 Marisita index (Mi). Within plots 1 and 2 used to calculate this index, 576 individuals were recorded, distributed as follows: plot one 289 individuals, plot two 287 individuals, which indicates a similarity between plots of 0.087%.

5.3.6.9 Sorensen's Coefficient of Similarity (.CS). The Sorensen's Coefficient of Similarity (.CS) was 24.306 %, i.e. there is a very low similarity between plots 1 and 2.

5.3.6.10 Statistical analysis and calculation of sampling error.

LINES	Sampled Area (ha)		Sampled Volume (m)3		Output (ha x m)3
	X	X2	Y	Y2	(X.Y)
1	0,5	0,25	54,361	2.955,118	27,181
2	0,5	0,25	50,895	2.590,301	25,448
3	0,5	0,25	48,248	2.327,870	24,124
4	0,5	0,25	61,870	3.827,897	30,935
TOTALS	2,00	1,00	215,374	11.701,186	107,687

Sampling fraction:

$$f=\frac{\sum x}{At}=\frac{2ha}{100ha}=0{,}02$$

Intensidad de muestreo:

$$\text{Im}=\frac{\sum x}{At}*100\%=\frac{2ha}{100ha}*100=2\%$$

Media del área muestreada

$$\overline{X}=\frac{\sum x}{n}=\frac{2ha}{4}=0{,}5ha$$

$$\overline{X}^2=0{,}25(ha)^2$$

Media del volumen muestreado

$$\overline{Y}=\frac{\sum y}{n}=\frac{215{,}374m^3}{4}=53{,}844m^3$$

$$\overline{Y}^2=2899{,}068(m^3)^2$$

Media de $\bar{x}$ del área muestreada por la media del volumen muestreado $\bar{y}$

$$\bar{x}*\bar{y}=53{,}844ha*0{,}5m^3=26{,}922ha*m^3$$

Volumen medio por hectárea

$$\bar{q}=\frac{\bar{y}}{\bar{x}}$$

$$\bar{q}=\frac{53{,}844(m^3)}{0{,}5(ha)}=107{,}686(m^3)/ha$$

$$\bar{q}^2=11596{,}274(m^3/ha)^2$$

Calculation of sampling error

$$S^2_{\bar{q}} = \frac{(1-f)}{n*(n-1)} * \bar{q}^2 * \left[\frac{\sum x^2}{\bar{x}^2} + \frac{\sum y^2}{\bar{y}^2} - 2\left(\frac{\sum x*y}{\bar{x}*\bar{y}} \right) \right]$$

$$S^2_{\bar{q}} = \frac{(1-0.02)}{4*(4-1)} * 11596{,}274(m^3/ha)^2 * \left[\frac{1ha^2}{0{,}25ha^2} + \frac{11701{,}186(m^3)^2}{2899{,}068(m^3)^2} - 2\left(\frac{107{,}687(m^3*ha)}{0{,}5ha*53{,}844m^3} \right) \right]$$

$$S^2_{\bar{q}} = 950{,}930(m^3/ha) * 0{,}036$$

$$S^2_{\bar{q}} = 34{,}233(m^3/ha)$$

Varianza: $S^2_{\bar{q}} = 1171{,}898(m^3/ha)$

<u>Desviación estándar</u>

$$S_{\bar{q}} = \sqrt{S^2_{\bar{q}}}$$

$$S_{\bar{q}} = \sqrt{34{,}233(m^3/ha)^2}$$

$$S_{\bar{q}} = 5{,}850(m^3/ha)$$

<u>Calculo del error de Muestreo</u>

$$E\% = S\bar{q}\% = \frac{\pm S\bar{q}*100}{\bar{q}}$$

$$E\% = \frac{5{,}850(m^3/ha)*100}{107{,}686(m^3)/ha}$$

$$E\% = 5{,}433\%$$

Studen's t test t5 % 8gl

$$L_1, L_2 = \bar{q} \pm t^{5\%}_{gl=7} * E\%$$

Confidence limits:

- Upper limit: L_1 = 107.686 m^3 /ha + 1.895 * 5.433 = 117.981 m /ha^3
- Lower Limit or Reliable Volume: L_2 = 107,686 m^3 / ha - 1,895 *5,433 = 97,390 m /ha.[3]
- The volume of the inventory is between: **117,981 m^3** and **97,390 m^3**

with the most reliable volume of **97,390 m^3**

❖ Volume of the whole forest 36.915,240 **m^3**

5.3.7.1 Volume per species and per line for all trees found with DBH >= 10 cm. A total volume of 738,305 m^3 was recorded, distributed within the four lines as follows: line four with a volume of 187,471 m^3 , line three with 187,926 m^3 , line two with 162,312 m^3 , and line one with 200,596 m.^{3}

5.3.7.2 Volume per species and per line for all trees found with DBH>10 cm and <= 35 cm. For this category, a volume of 353,759 m^3 was recorded, with the following lines being the most representative: line four with 99,983 m^3 and line three 90,818 m .^{3}

5.3.7.3 Volume per species and per line for all trees found with DBH>40 cm. A total volume of 392,271 m^3 was recorded, distributed within the four lines from largest to smallest as follows: line one 115,214 m^3 , line three with 97,102 m^3 , line four 93,708 m^3 , and line two 86,247 m .^{3}

5.3.7.4 Volume per species and per area sampled for all trees found with DBH >= 10 cm. An average volume per hectare of 369,152 m^3 and a total volume of the 100 ha sampled of 36.915,240 m^3 were recorded, where the most representative species were: Lechero (Brosimum utile) with an average volume of 57,248 m^3 and a total volume of 5.724,750 m^3 , Carrá (Humberodendro patinoi) with an average volume of 25,747 m^3 and a total volume of 2.574,650 m^3 , Boteco (Matisia sp) with an average volume of 16,355 m^3 and a total of 1.635,450 m^3 , Anime (Protium nervosum cuatr) with an average volume of 14,251 and a total of 1.425,100 m^3 ; among the least representative recorded were: Manteco (Pera arborea) with an average volume of 0,006 m^3 and a total volume of 0,550 m^3 , Guacharaco (N.N) with an average volume of 0.028 m^3 and a total of 2,800 m^3 and Castañeta (N.N) with an average volume of 0.034 m^3 and a total of 3,350 m .³

5.3.7.5 Volume by species and by area sampled for all trees found with DBH >10 cm and <= 35 cm. An average volume of 176,880 m^3 and a total volume of 17,687,950 m^3 were recorded, of which the most representative species were: Milkwood (Brosimum utile) with an average volume of 12,874 m^3 and a total volume of 1.287,400 m^3 , Ant (Lunania paruiflora spr.et Benth) with an average volume of 6,408 m^3 and a total of 640,750 m^3 , Boteco (Matisia sp) with an average volume of 6,201 and a total of 620,100 m^3 , Anime (Protium nervosum cuatr) with an average volume of 5,720 m^3 and a total of 571,950 m^3 ; among the least representative were:. Manteco (Pera arborea) with an average volume of 0.006 m^3 and a total volume of 0.550 m^3 , Guacharaco (N.N) with an average volume of 0.028 m^3 and a total of 2.800 m^3 and Castañeta (N.N) with an average volume of 0.034 m^3 and a total of 3.350 m .³

5.3.7.6 Volume by species and by area sampled for all trees found with DBH >=40 cm. A total of 54 individuals and an average volume of 196.136 m^3 and a total volume of 19,613.550 m^3 were recorded, distributed within the most representative species as follows: Lechero (Brosimum utile) with an average volume of 44,334 m^3 and a total volume of 4.437,,350 m^3 , Carra (Humberodendron patinoi) with an average volume of 21,459 m^3 and a total volume of 2.145,850 m^3 , Boteco (Matisia sp) with an average volume of 10,154 m^3 and a total volume of 1.015,350 m^3 , the least representative were: Chanó (Sacoglostis procera) with an average volume of 0.455 m^3 and a total volume of 45,500 m^3 , Lirio (Couma macrocarpa) with an average volume of 0.791 m^3 and Choibá (Dipteris panamensis) with an average volume of 0.791 m^3 and a total volume of 79.050 m .^{3}

5.3.7.7 Volume per line and per area sampled for all trees with DBH >= 10 cm. A sampled volume of 369,152 m^3 and a total volume of 36.915,240 m^3 were recorded, the most representative lines were: line one with a volume of 100,298 m^3 and a total of 10.029,800 m^3 , line three with an average volume of 93,963 m^3 and a total volume of 9.396,300 m.^{3}

5.3.7.8 Volume per line and per area sampled for all trees with DBH >=10 cm and <= 35 cm. For the latter, a volume of sampled of 176,880 m^3 and a total volume of 17,687,950 m^3 , distributed along the lines from most to least representative.

5.3.7.9 Volume per line and per area sampled for all trees with DBH >=40 cm. An average volume of 196,136 m^3 and a total volume of 19.613,550 m^3 were recorded, where the most representative lines were: line one with an average volume of 57,607 m^3 and a total volume of 5.760,700 m^3 and line three with an average volume of 48,551 m^3 and a total volume of 4.855,100m .^{3}

5.3.7.10 Volume per family and per line of species with DBH >= 10 cm. Thirty-four families were recorded with a total volume of 742,954 m^3 , distributed within the four lines. The most representative were: Moraceae with 122,943 m^3 , Bombacaceae with 95,850 m^3 , Lauraceae with 59,187 m^3 , the least representative were: Verbenaceae with 0,231 m^3 , Combretaceae with 1,286 m^3 , Simaurobaceae with 3,327 m .^{3}

5.3.8.1 Volume per family and per line of species with DBH >= 10 cm and <= 35 cm. A total volume of 450,403 m^3 was recorded, and the most representative families were: Moraceae with 69,293 m^3 , Bombacaceae with 48,289 m^3 , Lauraceae with 44,978 m^3 ; the least representative were: Verbenaceae with 0,231 m^3 , Celastraceae with 1,252 m^3 , Combretaceae with 1,286 m^3 respectively.

5.3.8.2 Volume per family and per line of species with DBH >= 40 cm. A total volume of 390,114 m^3 was obtained, with the following families being the most representative: Moraceae with 92,449 m^3 , Bombacaceae with 66,434 m^3 , Anacardiaceae with 23,622 m^3 , and the least representative were the following families representative were: Apocynaceae with 1,581 m^{3,} Annonaceae with 1,581 m^3 , Clusiaceae with 2,008 m^3 and Tiliaceae with 2,472 m .^{3}

5.3.8.3 CHARACTERISTIC OF NATURAL REGENERATION.

Floristic composition. Within the natural regeneration, 27 families, 77 species and 199 individuals were found.

The most representative families were: N.N with 20 individuals, Euphorbiaceae with 11 individuals, Lauraceae with 9 individuals, Burseraceae with 8 individuals and Myristicaceae, Aceraceae with 7 individuals respectively.

5.3.8.4 Density, mixing coefficient and abundance of natural regeneration for the categories Renuevo, Brinzal and Latizal. The most representative species were: Anime with 14 individuals and a relative abundance of 7.955 %, Incibe with 10 individuals and a relative abundance of 5.682 %, Otobo with 9 individuals and a relative abundance of 5.114 %, and the least representative were: fruta, Verbenaza, Rabo de iguana, Pampanillo, Palma táparo with 1 individual and a relative abundance of 0.568 %, respectively.

In the Brinzal category, a total of 157 individuals were recorded, where the most representative species were: Algodoncillo with 13 individuals and a relative abundance of 8,280 %, Palma sin rama with 9 and a relative abundance of 5,732 %, Palma meme, Quiribe with 8 individuals and a relative abundance of 5,096 %, Incibe with 7 individuals and a relative abundance of 4,459 %. And the least representative were: Mayorquín with 4 individuals and a relative abundance of 0.433 %, Parrillo, Pampanillo, Palma rabo de zorra with 1 individuals and a relative abundance of 0.637 %. The most representative species were: Sajo with 7 individuals and a relative abundance of 4.930 %, Aceite maria, Dinde with 6 and a relative abundance of 4.225 %, Algodoncillo, Costillo, Guamo with 5 individuals and a relative abundance of 3.521 %, and the least representative species were: Verbenaza, Vaina, Perdis, Parrillo, Palo blanco with 1 individual and a relative abundance of 0.704 %.

5.3.8.5 Frequency of natural regeneration for the categories Renuevo, Brinzal and Latizal. In the category Renuevo a total of 92 individuals were recorded, where the most representative species were: Carra, Incibe with 4 individuals and a relative frequency of 4.348 %, Anime, Caimito de monte, Chanó, Choibá with 3 individuals and a relative frequency of 3.261 %, and the least representative were: Virgusa, Verbenaza, Vaina, Uva, Tuabe with 1 individual and a relative frequency of 1.087 %, each one.

In the Brinzal category, a total of 157 individuals were recorded where the most representative species were: Lechero with 8 individuals and a relative frequency of 5.096 %, Nuánamo. White Ant, Caimito and Guamo with 6 individuals and a relative frequency of 3.822 %, Algarrobo, Caraño, and Anime with 5 individuals and a relative frequency of 3.185 %, and the least representative were: Guamo amarillo, Uva, Jigua amarillo, Castaño, Pantano, Tometo, Hueso and Flor de rosa with 1 individual and a relative frequency of 0.637 %, respectively. In the Latizal category, a total of 131 individuals were recorded where the most representative species were: Hormigo blanco with 6 individuals and a relative frequency of 4,580 %, Guamo with 5 individuals and a relative frequency of 3,817 %, Bolenillo, Anime, Carbonero, Hormigo colorado, Jigua negro with 4 individuals and a relative frequency of 3,053 %, respectively, and the least representative were: Guayacán negro, Casaco, Pampanillo, Laurel, Castaño with 2 individuals and a relative frequency of 1.527 %, respectively , Uva, Cedro macho, Tometo, Flor de rosa, Bongo, Algodoncillo and Carate with one individual and a relative frequency of 0.763 %.

5.3.8.6 Criteria for the selection of harvestable species. The most important criteria to be taken into account for the selection of the species to be harvested is that it has the minimum cutting diameter required by the autonomous corporation of each region, which in our area is 40cm DBH.

6. ENVIRONMENTAL CONSIDERATIONS

6.1 MEASURES TO PREVENT AND MITIGATE IMPACTS ON BIOTIC AND ABIOTIC RESOURCES.

The measures that must be taken into account within the inventory to avoid negative environmental impacts, or to mitigate the use of these species are mainly based on good silvicultural management that minimises the deterioration of the biotic and abiotic components generated by this type of work and at the same time guarantees the sustainability of the forest and the resources that exist in it.

6.1.1 Environmental Considerations in Forest Harvesting. In general, forest resources will be extracted in an artisanal way, with primary techniques that destroy the vegetation that would later replace the adult forest, cutting down easily accessible forest along rivers and streams. Cutting techniques are inefficient and a high percentage of the material that is cut remains in the forest and is not incorporated into the markets.
In order to make a good use of the forest mass without causing such a negative environmental impact, the selection of the trees to be felled, as well as the different techniques to be used, must be taken into account. The forest management plan should also include a good management plan that reduces the negative effects on the environment caused by the harvesting of the forest and determines how to compensate for the damage caused to the ecosystems present in the forest.

6.1.1.1 Access routes (major transport). Due to the topographic conditions in the area and the difficulty that may arise from transporting the timber by land, it is considered that the best way to transport the timber to the processing site would be by water due to the conditions in the village and the difficulty that would arise from the use of other means of transport.

6.1.1.2 Tree felling work. Tree felling is an activity that is carried out to move the tree from a vertical to a horizontal position; in reality, these tasks are carried out using tools such as axes and chainsaws. The latter has allowed a greater extraction of the resource, as it is pushed by fuel. In general, the felling work in the Pie de Pepé district is carried out at the beginning of the summer season, to guarantee greater mobilisation of the wood, as this facilitates the task of the mules.

6.1.1.3 Log stockpiling. Once the logs have been sawn and transported to the marketing site (Istmina), they are stored and marketed, in many cases quickly.

6.1.2 Biodiversity Conservation. Pie de Pepé is a village with a great variety of plant and animal species. It is essential to generate management plans that guarantee the persistence and sustainability of the resources.

6.1.2.1 Impact mitigation measures and increased benefits. There is a need to implement measures that ensure that negative impacts are minimised and that in fact provide greater socio-economic benefits to producers in the area, such as the use of silvicultural forest management techniques in the event that the forest can be renewed or reforested, but cannot regenerate itself in a sustainable way to ensure economic income and therefore a better way of life for the community.

6.2 SOIL AND WATER CONSERVATION.

6.2.1 Soil Resource. They are limited by drainage (waterlogging) due to the presence of rubble and/or stones, or by frequent flooding, and therefore require a series of adaptation and management practices in accordance with their nature. It is important to conserve these soils because they can be used, mainly for intensive crops, and they can be used for forestry, due to the fact that their ecosystems have a high diversity of flora and commercial timber species.

These soils are favoured by the flat topography and an optimum pH that allows them to provide nutritional elements to the plants without major restrictions. The land has an agricultural vocation for products such as bananas, borojo, chontaduro, bananas and fruit trees.

6.2.1.1 Expected impacts. These soils are characteristic of the very humid tropical regions of the country; they are lands of important forestry vocation in the balance of the ecosystems of the humid tropics.

They correspond to flat and hilly sectors, but dedicated to agricultural activity with adverse factors such as shallow soils, materials susceptible to erosion, arrangement of the strata, which naturally facilitate erosion.

6.2.1.2 Measures to mitigate impacts and increase benefits. The aim is to implement forestry and silvicultural management measures and techniques that provide greater socio-economic, environmental and ecological benefits to producers and the population in general.

6.2.2 Water Resources. The water resource is very abundant in the Pie de Pepé district, it is used as a means of transport and for the supply of drinking water and as a means of subsistence through artisanal fishing. This implies, therefore, an adequate management of this resource for the sake of its conservation. The vulnerability of the rivers and streams of the corregimiento is due to the increasing contamination derived from the discharge of sewage, excreta and rubbish. The intensities of these practices affect the water bodies despite the fact that they have their own cleaning mechanisms.

6.2.2.1 Expected impacts. Water resources can constitute hazards because they are a potential danger to crops, pastures and the resident population. Floods occur when intense downpours or heavy rainfall Floods occur in flood plains, flood plains, flood plains, flood plains, flood plains, flood plains, flood plains, flood plains, flood plains and flood plains. Flooding occurs in flood plains, flood plains, specifically in river valleys and low terraces, and is maximised when the vegetation cover that regulates the water regime has disappeared or has been drastically reduced. Floods constitute a threat when the aforementioned areas are used for purposes other than protection, causing economic and human losses.

6.3USE OF CHEMICALS.

6.3.1 Handling of fuel and lubricants. The use of tools is indispensable in a harvesting process. These tools are based on fuels and oils, which must be located in clean and safe places in order to avoid losses and spreading in the area. On the other hand, special care must be taken when using oils and fuels, as they have a negative influence on the water when extracting wood from the forest, and if they are not well managed, they can contaminate the water and cause damage to the species in the stream.

6.3.2First Aid. When entering the forest or the study area, a first aid kit should be carried, with medicines that are easy to handle and quickly effective, in case of an accident. The purpose of this is not to suspend work and thus avoid wasting time in the course of carrying out the different activities involved in forest harvesting.

6.3.3Fire. Although forest fires are not so common in this environment, either due to the sun or high temperatures, great care must be taken to maintain a good vegetation cover in order to preserve the relative humidity of the soil to avoid drying out and wilting of the smaller plants. In addition, no plastic bags or other litter should be thrown in the forest.

6.4WASTE MANAGEMENT.

In this district there is no sewerage system, but most of the population has a sanitary toilet as an individual solution, the water is evacuated through pipes to the Banderillero, El Bracito and Pepé river streams. The discharge of sewage affects the water sources, their environmental stability given the number of people and the location of the discharge, which corresponds to the upper part or source of the river Pepé. It should also be taken into account that populations downstream are served by the same source. The coverage of the sanitation service or collection and treatment of solid waste is 0% in the whole municipality. The absence of this service is quite noticeable, most of the families of the corregimiento throw the rubbish in the back of the houses, commonly called "Paleadera", allowing the proliferation of some insects (flies and mosquitoes).There is also no place designated exclusively for the deposit of solid waste. The majority of the inhabitants dump their waste directly in the waste into rivers or streams. Another type of evacuation that occurs is the temporary deposit of waste in backyards for later disposal in the river. The absence of land communication routes makes it difficult to collect and deposit waste centrally.

6.5ENVIRONMENTAL CONSIDERATIONS IN HUMAN HEALTH.

The supply of drinking water is one of the fundamental priorities as one of the basic services. It requires strategic planning based on criteria of effectiveness, efficiency and social impact. It is necessary to provide a complete water supply system, either by gravity or by pumping (intake, tanks and distribution network). The village requires different types of alternatives for the management of sewage and solid waste. In order to improve human health conditions in this population, it is suggested to establish sewage systems with a complete network of collectors, wells and aeration lagoons; also a system of biodigester septic tanks as an alternative wastewater treatment system, including the provision of multi-family septic tanks and collectors; and to provide a sanitary landfill system, with certain proximity to the urban area, creating additionally a solid waste collection system adequate to the characteristics of these populations.

6.6 STRATEGIES AND INSTRUMENTS FOR ENVIRONMENTAL MONITORING BY SUSTAINABLE FOREST MANAGEMENT STAKEHOLDERS.

In this aspect, it is proposed that natural wealth should be a real option for development, based on actions such as the promotion of the sustainable use of natural resources (forest, fauna, soil) to achieve food security, given the ancestral productive practices linked to the natural environment. Promote the value of environmental services, the advanced uses of biodiversity and the rights of communities over them, and the strengthening of research and appropriate technological supply. The forest must be managed sustainably, in order to obtain resources for the present generation and guarantee the resource for future generations. In this sense, the Autonomous Corporations are responsible for overseeing this strategy.

6.6.1 For the Environmental Authority (CODECHOCÓ). The Corporación Autónoma Regional para el Desarrollo Sostenible del Choco (CODECHOCÓ), is the environmental authority in charge of ensuring that sustainable development is achieved and maintained. This term is applied to economic and social development that meets the needs of the present without compromising the ability of future generations to meet their own needs. There are two fundamental concepts regarding the sustainable use and management of natural resources. First and foremost, the basic needs of mankind (food, clothing, shelter and work) must be met. The main objective of achieving sustainable development for a community should be to ensure food security to meet all basic needs. The aim is to create a high quality of life for the population as a whole.

6.6.2 For the community. To make the community of Pie de Pepé aware of the need to adopt a system of sustainable forest use and management of renewable natural resources that will improve their social, economic, ecological and technical standard of living.

7. COMMUNITY PARTICIPATION

With regard to the organisation and participation of the community in decision-making processes and actions for socio-economic development, it can be established that the community of Pie de Pepé is very organised; there are traditional practices of solidarity and compadrazgo, in projects of common interest. This community has a social structure organised around cultural and development projects in general.

BIBLIOGRAPHY

Esquema de Ordenamiento Territorial Medio Baudó (2006-2016). CODECHOCO - IIAP. Plan de Ordenación Territorial Medio Baudó (2016). CODECHOCO - IIAP.

Forestry Technician (2000). First Edition.

Printed by Books on Demand GmbH, Norderstedt / Germany